The *Real* Origin of the Species

Twelve New and Compelling Reasons to Believe That God Exists

by

Oscar J. Daniels, Sr.

TEACH Services, Inc.
Ringgold, Georgia

**PRINTED IN
THE UNITED STATES OF AMERICA**

World rights reserved. This book or any portion thereof may not be copied or reproduced in any form or manner whatever, except as provided by law, without the written permission of the publisher, except by a reviewer who may quote brief passages in a review.

The author assumes full responsibility for the accuracy of all facts and quotations as cited in this book.

Copyright © 2010 TEACH Services, Inc.
ISBN-13: 978-1-57258-632-1
Library of Congress Control Number: 2010937129

Protected by copyright and by registration with the Writers' Guild of America.

Published by
TEACH Services, Inc.
www.TEACHServices.com

Contents

Foreword v

Interdependency 1

Evolution or Creation? 11

When Did Life on Earth Begin? 25

The Holy Bible: Is it Fact or Fiction? Now it is Possible to Know for Sure! 33

    Section I: The Biblical Account of Creation Compared to the Theory of Evolution 33

    Section II: Follow the Logic 44

    Section III: Is Evolution a Form of Idolatry? 47

    Section IV: The DNA Evidence is in, and a Positive Identification Has Been Made: God is the Real Father of Us All 51

    Section V: Positive Proof of the Veracity of the Bible: A Look at Prophecy and History 61

    Section VI: Implications and Recommendations 67

Foreword

There is much confusion today about how the world began and about the origin of life on earth. Schools and universities teach one thing, but churches and theologians teach something altogether different. How can we know with absolute certainty who is right? This book enters the debate with some surprising and compelling arguments. By the time you finish reading this fascinating material, you will no longer be in doubt. For you, the debate will be over. You will not only be more confident of your position on these issues, but you will also have some new and more effective ways to defend and share your faith.

Chapter 1

Interdependency

Did life on earth evolve over millions of years as scientists claim, or was it created in one week as the Bible says? The answer to that question can be found in a concept that I call "Interdependency". It's a new and closer look at the science of ecology. Ecology is the study of how various things in nature depend on each other in order to survive or function properly.

There are more than 1.8 million different species of insects, birds, and animals. All of these species have one thing in common: they all require that there be a male and a female in order to reproduce. Reproduction is the only way to ensure the survival of each and every species.

The Real Origin of the Species

What are the chances that a male and a female would evolve at the very same time? There is no chance at all. The simultaneous production of a male and female would never occur with the slow and haphazard process of evolution. If the process of evolution did occasionally produce a male and female at the same time, it would be truly amazing; but to claim that it happened 1.8 million times among all of the various species is simply not believable. The male and female in all of the different species are clearly designed for each other in every conceivable way. Not only were they all perfectly matched, but they all came into existence at the same time as their mates. This allowed them to begin reproducing immediately, which ensured the survival of each of the various species. Those kinds of things don't just happen by chance or by some sort of evolutionary accident. They only happen by design; and where there is a design and a purpose for anything, there has to be a designer.

Take a single species like human beings, for example. If it took millions and millions of years for the first man to evolve, would he have then

Interdependency

had to sit around and wait millions more years for his female counterpart to evolve? None of the myriad species of living creatures could have survived in a slow, random, and uncontrolled process like that.

The need for a male and female to be immediately available in every species is only one example of interdependency. All living creatures have very specific requirements as far as habitat and food supply are concerned, and all of these requirements had to be met on the very first day of their existence. This required careful planning and preparation by an intelligent designer, because none of the various life forms could have waited long periods of time for their habitat and food supply to evolve.

All living creatures are dependent on each other in countless ways other than the need for reproduction. For example, flowers require bees and other insects for cross-pollination, and the insects require the plants for food. If the plants evolved first and then had to wait millions of years for the insects to evolve, they would not have survived. The opposite is also true. If those insects had evolved first, they would have starved to death waiting for plants to evolve. However, since God created the insects on the day after the plants were created, the timing was perfect in order to supply the needs of both.

At the present time, there are approximately one million insects per acre of land. They multiply rapidly, and if they had evolved millions of years before birds, bats, and other insectivores, the entire earth would have been inundated with them. On the other hand, if the animals that feed on insects had evolved millions of years before the insects did, those animals would have starved to death waiting for their food supply to show up. They both had to come into existence at the same time, and not millions of years apart. Since insects and insectivores were created only one day apart, the timing was perfect in order to maintain the proper ecological balance right from the start. This balance among the various species would not have been possible if they had evolved millions of years apart.

Interdependency

The entire food chain requires the simultaneous existence of all living creatures. Polar bears eat seals; seals eat fish; and fish eat smaller creatures and plants. Lions and tigers eat deer, antelope, and oxen. Wolves, foxes, and birds of prey eat rodents and other small animals. None of these animals could have survived unless their food supply was immediately available.

There are thousands of examples of these interdependencies, and I could go on identifying them, but I think my point has been made. The proper functioning of the earth's ecosystems does not lend itself to long periods of time lasting millions of years. The various components of the ecosystem could not have survived during long evolutionary periods while waiting for all of the other essential components to evolve. All of these components had to be available immediately, as occurred in the biblical model.

Everything on earth has a specific purpose—a purpose that fits neatly into an overall master plan. The fact that everything has a purpose is still further evidence of God's existence. Only a sentient being is able to work with a sense of purpose, and then design things to fulfill that very purpose. Since the so-called process of evolution is not a living entity, it has no conscious awareness of what it is doing. Without that kind of awareness, the process of evolution cannot work from a master plan, nor can it produce things that are designed to fit into the grand scheme of things. Without any kind of consciousness or sense of purpose, evolution could never create the beauty, harmony, and functionality that we see everywhere all throughout earth's vast ecosystems. Without having a conscious awareness of the future, evolution could never anticipate the special needs of all the plants and animals, nor could it provide for those needs in advance. Only someone with God's powers and awareness could do those things. Without conscious awareness, there is only chaos.

Not only did the Lord establish a purpose for everything on this planet, but he also told all of his creatures what that purpose was. The Bible says that on the very first day of Adam and Eve's existence, God told them what they were

Interdependency

supposed to do, what they were supposed to eat, and where they were supposed to live. Since all of the other animals knew these same things on the very first day of their existence, we can assume that God gave them this knowledge as well. If the millions of animals on this planet had come into existence by the process of evolution, they would not have had the benefit of these instructions. Instead, they would have had to figure these things out by trial and error over an extended period of time. No doubt most of them would have starved to death and become extinct almost immediately.

Those who believe in the Theory of Evolution should ask themselves the following questions:

1. How is it possible that on the very first day of their existence, all 1.8 million different species found their own uniquely suitable habitat and food supply already waiting for them? Not only did they find these things waiting for them, they also knew exactly what they were supposed to eat and where they were supposed to live. They could not

have taken weeks, months, or some longer period of time to figure out these things on their own. They would never have survived.

2. The next question to ask yourself is this: How could the process of evolution produce a magnificent world like ours when it is not a living entity and has no conscious awareness of what it is doing?

3. How is it possible that on the very first day of their existence, each one of those 1.8 million different species found a compatible mate, properly designed and equipped for reproduction and for feeding their offspring? Did that all happen by some very fortunate coincidence, or did it happen as a result of careful planning by an intelligent designer? The only logical conclusion is that it happened by design and within a brief period of time, just as the Bible says. The order, beauty, harmony, and interdependencies that exist within the different ecosystems and throughout the universe all pro-

Interdependency

claim themselves to be the handiwork of an omnipotent and omniscient creator.

Ask yourself those three questions, and see what conclusion you reach. The odds of all those things happening by themselves are beyond calculation. It simply could not have happened on its own.

Adam and Eve in their garden home

Fortunately, Adam didn't have to mope around by himself while waiting for Eve to evolve. God

The Real Origin of the Species

provided her on the same day that Adam was created, just like he provided our food supply and all of our other needs. This didn't happen over millions of years, but in a timely manner.

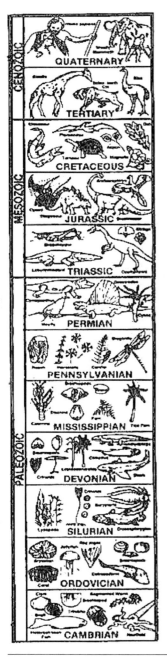

Chapter 2

Evolution or Creation?

A Look at What the Geological Column Reveals

The Geological Column is a diagram that shows the various layers that comprise the surface of the earth. It also shows the different animals whose remains have been found in each of those layers. Scientists who study the Geological Column have reached some conclusions that are at odds

The Real Origin of the Species

with what the Bible says. Which one is correct? Let's take a look at the evidence, and then you can decide for yourself.

Scientists are quick to point out that distinctly different kinds of animals are found in each of the layers that form the surface of the earth. They claim that these layers represent different geological ages lasting millions of years, and they cite this fact as proof that the animals in those layers evolved separately and sequentially over many millions of years.

The only problem with this conclusion is the fact that it's based on a false premise. The sedimentary layers in which those animals are found were not formed over millions and millions of years. They were, instead, formed rapidly in one year during the great flood mentioned in the Bible.

There are several factors that support this position:

1. Sedimentary layers are only produced by water. Since the entire surface of the earth

Evolution or Creation?

is covered with sedimentary layers containing fossils, this is proof that the entire earth was once under water at a time subsequent to the appearance of animals. This agrees with the biblical account of the Flood in Noah's day.

2. Buried in those sedimentary layers are whole forests full of petrified trees that are still standing upright. In the normal course of events, trees rot and decay; and then after a while, they fall over and are devoured by termites. They do not remain standing while sediment slowly builds up around them over millions of years. If the sedimentary layers had been formed over millions of years as scientists claim, there would be no standing trees within those layers.

Trees do not live or remain standing for millions of years. How do we know that they don't live that long? We know this because trees produce annual rings inside their trunks each year, and no one has ever seen a tree with a million annual rings. So, the fact that those trees are still standing is evidence that the sedimentary layers were formed rapidly, as would occur during a flood. They were not formed over millions of years as scientists claim. These layers were formed so quickly that the trees were completely buried before they had time to rot and fall over.

There are distinctly different categories of plants and animals in each one of the sedimentary layers, and scientists claim that this is proof of the order or sequence in which they evolved. Once again, this is an erroneous conclusion. This is not evidence of the order in which those plants and animals evolved. It is, instead, further evidence of the biblical Flood.

If you look at the Geological Column, you will notice that there are twelve layers of sediment. In the third layer from the bottom, you will begin to see such things as algae, kelp, and other kinds of seaweed. In the fourth and fifth layers from the bottom, you will begin to see trees and other kinds of land-based plants. You don't have to be a rocket scientist to figure out that the algae,

Evolution or Creation?

kelp, and seaweed were already occupying some of the lowest places on earth when the Flood began. Since they were already so far below sea level to begin with, it stands to reason that they would be buried in the lower layers of sediment. Similarly, trees and plants were rooted in the soil just above sea level, and they would have been buried in the next few layers of sediment. The location of these plants in their respective layers of sediment had absolutely nothing to do with their evolving millions of years before the animals. They were simply trapped at some of the lowest places on earth, unable to escape to higher ground like the animals that were buried in the upper layers.

Another thing that helped to determine the layers in which the various animals were found was a naturally occurring process called "Sorting by Weight", and it is based on two laws of physics: the law of gravity and the law of buoyancy.

In a flood situation, where everything is swirling around in the water, objects are going to be sorted or separated by weight, with the heaviest

The Real Origin of the Species

objects settling to the bottom first. This explains why dinosaurs are found in the lower layers. They are in those layers simply because they were the largest and heaviest creatures on earth, and not because they evolved first, as scientists erroneously claim. Lighter-weight animals are sorted out in the upper layers based on a number of factors, including their ability to swim for a while and their ability to reach higher ground before succumbing to the floodwaters.

Since fish can swim, couldn't they have survived the Flood? Wouldn't they have been found in the upper layers of sediment instead of the bottom one? The answer is that they could not have survived, and the reasons are threefold: (1) their food supply was below sea level, and they could not live very long without it; (2) all of the gritty sediment floating around in the water clogged their gills, making it impossible for them to breathe; and (3) the fish were accustomed to a saltwater environment, and the sudden addition of so much fresh rain water diluted the oceans in which they lived. This abrupt and dramatic change in their environment no doubt hastened

Evolution or Creation?

their demise. For these reasons, they died during the early stages of the flood, and are buried in the bottom layer of sediment.

Whales, on the other hand, lasted a bit longer than the fish because they were able to go to the surface and breathe fresh air for a while. As a result, they are found in some of the middle layers of sediment instead of at the bottom with the fish.

Birds are in the middle and upper sedimentary layers, not because they evolved at that particular time, but because of their ability to fly to some of the highest points on earth for temporary safety. Unfortunately, birds depend on plants and insects for food; and when the Flood swept away their food supply, they could not survive, even at the high points where they sought refuge. In addition, they don't weigh very much, and their partially hollow skeletons allowed their dead bodies to float around on the surface of the water before finally settling into their respective layers of sediment. All of these factors helped to determine the sedimentary layers in which the birds were found.

There were many other factors that played a part in determining the layers in which the various animals were found. There were factors such as the instinct for survival that all animals have, and there was also mankind's own ingenuity. Humans, for example, are in the upper layers of sediment, not because they suddenly emerged at the end of an evolutionary cycle, but because they were able to escape to higher ground or to climb aboard boats, rafts, or other floating objects. By these means, they were able to ride out the early days of the Flood until they either ran out of food or died from exposure.

The Geological Column shows that clams, oysters and various crustaceans are found in the very bottom layer, and scientists take that as evidence that they were the very first life forms to evolve. The ability of scientists to repeatedly misinterpret the evidence is both astounding and baffling. Even in the face of evidence to the contrary, they instinctively and consistently

Evolution or Creation?

interpret the data in a manner that supports their established model, which is based on evolution.

Since clams, oysters, and crustaceans have the lowest body weight of all the animals, their presence in the bottom layer of sediment might seem to contradict the principle I mentioned earlier when I said that the heaviest objects would settle to the bottom first. Since these creatures are lighter in weight than all of the other animals, shouldn't they be in the top layer? The answer to that question is easy. These creatures were not affected by the sorting process that separated the other animals because they were already living on the bottom of the ocean when the Flood began. Therefore, they were buried immediately by the first layer of sediment.

All of the factors I described in the previous paragraphs determined the layers in which the various life-forms were found. The process of evolution played absolutely no part in the matter.

Genealogical records in the Bible suggest that the earth is slightly more than 6,000 years old.

Scientists, however, rely on carbon dating to determine the age of various things. Using that technique, they concluded that the layers in the Geological Column, as well as the fossils contained therein, are millions of years old. There is a problem with that conclusion, however, because carbon dating is an inexact science. Time and time again it has been proven to be unreliable. Just how unreliable carbon dating is was revealed some time ago when the remains of a woolly mammoth were discovered in Russia. Scientists were called in, and they used carbon dating to determine how old it was. They concluded that it was a certain age; but when they tested another part of the very same animal, they found it to be thousands of years older. Different parts of the same animal had different ages. That example and others like it clearly demonstrate that carbon dating is not a reliable tool to use in determining the age of the earth, the age of fossils, or the age of anything else for that matter. This is especially true if the item being examined is more than a few thousand years old.

Evolution or Creation?

Another way that scientists determine the age of things is by taking core samples from glaciers and other large masses of ice. They have concluded that each layer of ice represents one year's accumulation, much like the annual rings in trees reveal their age. This, however, is a false assumption. The layers of ice are not formed annually, but are instead formed each time it rains. Based on their own incorrect assumption, scientists have concluded that some masses of ice are at least 75,000 years old. This conclusion is not supported by the facts.

Some basic mathematics will show how erroneous it is for scientists to claim that each layer of ice represents one full year. Assuming that it rains a minimum of twelve times a year in those arctic regions, and that each layer of ice represents a single rainfall, then the age of the ice mass would be 6,000 years old, which agrees with the biblical age of the earth (6,000 years x 12 rainfalls per year = 72,000 layers of ice).
These ice masses are either 6,000 years old or 72,000 years old, and it all depends on how you interpret the data. Scientists will instinctively

interpret the data in a way that will support the Theory of Evolution. How can we know for certain which interpretation is right? The answer to this question is best illustrated by an incident that occurred during World War II. At that time, a squadron of military pilots was flying over Greenland on a secret mission. Because of miserable weather conditions and extremely cold temperatures, their planes malfunctioned, and they were forced to land in that frozen region. They became famously known as "The Lost Squadron".

The pilots were eventually found and rescued, but their planes had to be abandoned. Years later, when modern technology made it possible to find the location of the planes, two Americans, Patrick Epps and Richard Taylor, went to the site. These men were intent on salvaging the aircraft, and they thought it would be a relatively easy task. However, they discovered that the planes were buried under many layers of ice. By scientific calculations, the ice above those planes would have been 600 years old, but the fact that the planes had only been there since

World War II was irrefutable proof that the scientific calculations were wrong. The ice above those planes was less than 50 years old.

Scientists always interpret the data in ways that make it fit the template of evolution. This practice leads them to false conclusions and puts them at odds with the biblical record. God continually confounds the wisdom of man.

The Bible has repeatedly been proven to be true and accurate in spite of man's relentless efforts to dismiss it as unscientific. We cannot trust any of the claims made by scientists regarding the age of the earth or the origin of life. All of their claims are based solely on speculation and theory. There is no scientific proof for any of it.

Jesus said, "Seek, and ye shall find" (Matt. 7:7). It's exciting to know that the truth can be discovered if we search for it diligently. The surest way to arrive at the truth is to begin with the premise that the Bible is true, and then look for evidence that will either prove or disprove your

The Real Origin of the Species

premise. You will find that the evidence supports the biblical record every single time.

Note: I would like to acknowledge the pioneering work of Dr. Walter Brown in developing the concept of "sorting by weight". This concept explains the sequencing of the animals in the Geological Column. With Dr. Brown's permission, I have written this chapter based on his concept. Dr. Brown's work on this subject is fully described in his book titled *In the Beginning*. I encourage you to obtain a copy of it for more information on this topic. Copies of the book may be obtained from the Center for Scientific Creation, 5612 N. 20th Place, Phoenix, AZ 85016. You can also visit his Web site at http://www.creationscience.com/onlinebook/ or call 602-955-7663.

The picture of the Geological Column at the beginning of this chapter comes from chapter 3, section II, of a book written by Dr. Gary Parker titled *Creation: Facts of Life*. Dr. Parker's book was published in 2006 by Master Books, a division of New Leaf Publishing Group in Green Forest, AR. The illustration is used with permission of the publisher.

Chapter 3

When Did Life on Earth Begin?

There is wide disagreement between theologians and the scientific community about when life on earth began. Theologians believe that life on earth began slightly more than 6,000 years ago. Scientists, however, believe that dinosaurs appeared approximately 230 million years ago, and that humans appeared 200,000 years ago (the rest of the animals are said to have appeared during the intervening timeframe with each group of animals separated by millions and millions of years). Whom should we believe? They can't both be right.

Theologians base their conclusions on genealogical records in the Bible, which support the notion of a relatively young earth. So how did scientists arrive at such astronomically large figures? They did it by using a technique called carbon dating. This method of dating is based on the rate of decay (or half life) of the carbon-14 isotope.

If carbon dating is used on some relatively young organic matter (like a 100-year-old human skeleton, for example), then the results of the carbon dating will be fairly accurate and can be confirmed with DNA and a birth certificate. Carbon dating has also accurately determined the age of ancient Egyptian remains that go back as far as the days of the pharoahs and the pyramids, and this can be confirmed with historical records.

Reassured by that kind of success with relatively young organic matter, scientists have concluded that carbon dating can be used to determine the age of everything, regardless of how old it is. They have made the assumption that the rate of decay is constant throughout all time and under all circumstances; but what if that assumption is not correct? What if there were some unusual phenomenon that caused a temporary accelera-

When Did Life on Earth Begin?

tion in the rate of decay? If that happened, it would throw calculations way off, and cause scientists to conclude that things were millions of years old when they were actually only thousands of years old. What are some of the things that could cause an acceleration in the rate of decay? Several possibilities come to mind:

1. There may have been some sort of cosmic event that occurred during prehistoric times. Cosmic events are often cataclysmic in nature, and can occur in a number of different ways. They can occur in the form of massive meteor showers which can have the same effect as thousands of atomic bombs. There is evidence at a site in Russia that meteors not only impact the earth physically, but they sometimes release a powerful and very destructive kind of radiation. Cosmic events can also occur when a group of stars go nova or super nova in our part of the universe. Another type of cosmic event is the solar flare. If there were some unusually large and sustained solar flares, they may have occasionally altered the rate of decay

throughout the entire solar system. These or some similar cosmic events could have caused a temporary spike in the rate of decay, and that would make things appear to be older than they really are.

2. Another possibility is that some unusual phenomenon occurred on earth that would have affected all previously existing matter. The one major cataclysmic event that occurred at just the right time in earth's history and might have had such a profound effect is the Flood of Noah's day. It is possible that the fossils which scientists claim are millions of years old experienced a temporary acceleration in their rate of decay because they were under water for more than a year during the Flood. This possibility could be what has skewed scientific calculations and caused them to be so far off.

Since water from the Flood was higher than the mountaintops all over the world, it undoubtedly contained sulfuric acid from volcanoes, as well as other kinds of acids, gases, and chemicals from various sources. Those acids would have had a profound effect on both organic and inorganic matter that existed before the Flood.

3. A third possibility is that the decay rate was temporarily accelerated by a combination of factors. Not only were the fossils in acidic water for a whole year during the Flood, but they were also under extreme pressure from the sediment in which they were buried and from all the billions of gallons of water that was on top of them. The combination of pressure and acids may have affected the rate of decay.

4. The Bible says that God spoke; and at his command, the universe came into existence. Although it was God's word that initiated creation of the universe, the process by which it occurred apparently involved a huge expenditure of energy. That powerful burst of energy produced a tremendous amount of heat and pressure. There are two

The Real Origin of the Species

things that provide evidence of this expenditure of energy:

A. The core of many planets and stars still contain hot, molten matter some 6,000 years after they were created.

B. It has been discovered that the universe is continuing to expand, and the rate of that expansion is accelerating instead of decreasing (this is not what one would expect after an ordinary and uncontrolled "Big Bang").

So, as a result of the initial heat and pressure associated with creation, we have planets and stars that are still hot and a universe that is still expanding. Every beginning chemistry student knows that heat and pressure accelerate chemical reactions and speedup the transfer of electrons, but scientists don't take that into consideration as they confidently construct their concepts on the shaky foundation of evolution.

When Did Life on Earth Begin?

It is possible that the initial burst of energy, along with the resultant heat and pressure, may have caused a temporary acceleration in the rate of decay for the carbon-14 isotope. After things settled down and cooled off a bit, the rate of decay may have slowed to its current rate. This would explain why rock samples from Mars, the moon, and Earth itself all appear to be billions of years old when they are actually only thousands of years old.

5. This final possibility is the one I find most intriguing. It addresses the long-standing controversy between scientists and theologians over the question of when life on earth began, and it does so in such a unique way that it just may be able to reconcile the differences between their opposing viewpoints.

The world that existed before the Flood was a very different place than the world that emerged afterward. In the beginning, it was not God's plan that people should die. Even after sin entered the world, people still continued for a while to live well over 900 years. Perhaps the physiology which made that great longevity possible also required an accelerated half-life for the carbon-14 isotope, and only after the Flood did it gradually slow down to its current rate of decay.

In this scenario, the cells in the human body might have been replaced at a faster rate than they are today, thereby keeping the body eternally young and vibrant. That would have had the effect of making all organic matter that existed before the Flood appear to be millions of years old when it was actually only thousands of years old.

On the issue of carbon dating, God has somehow managed to confound the wisdom of the wise, and to bring to nothing their understanding of this matter. We will probably never know for certain exactly what factors have skewed the scientific calculations so dramatically. We can only speculate about it, as I have done here; but if I have to choose between trusting the word of God or the word of scientists, I'll choose the word of God every time.

Chapter 4

The Holy Bible: Is it Fact or Fiction? Now it is Possible to Know for Sure!

Section I

The Biblical Account of Creation
Compared to the Theory of Evolution

From cover to cover, the Bible is full of fantastic stories. Some of them are so amazing that they stagger the imagination. Are they true? Did these things really happen, or are they merely a collection of myths, legends, and allegories designed to inspire the faithful? This book will at-

The Real Origin of the Species

tempt to answer these questions, not with opinions, but with solid and compelling evidence.

The very first verse in the Bible starts with one of those fantastic stories. It declares that God spoke; and at his command, the entire world and all living things came into existence. All of this happened within just six short days, approximately 6,000 years ago. Wow! That just doesn't seem possible, does it? What rational person could possibly believe a story like that? After all, we have brilliant scientists who, with great authority and certainty, assure us that the world began quite differently. So, just for the sake of comparison, let's take a look at the scientific explanation of how everything began, and see which version makes more sense.

Scientists claim that the world was formed several billions of years ago when a tremendous explosion occurred. They don't give any explanation of what caused the explosion or the source of the physical matter involved in it. Known as "The Big Bang," this explosion was so powerful that it flung trillions of gigantic chunks of matter

The Holy Bible: Is it Fact or Fiction?

into space; and somehow, it mysteriously organized itself into various solar and planetary systems that began to follow precise laws of gravity and motion. This cosmic space debris fell into orderly and highly predictable orbits. We are told that this process was completely random, and that it occurred all by itself.

After things settled down in the universe, one little chunk of space debris, which would later be known as the planet Earth, got caught in the gravitational pull of the sun. Miraculously, it began orbiting around the sun at a safe and precise distance (not too close, lest it burn up; and not too far away, lest it become frozen and barren). It all just kind of happened on its own and without any supernatural design or involvement. It's truly amazing how many things there are that can happen all by themselves.

What's even more amazing is the fact that this little chunk of space debris, which had no particular purpose, mysteriously acquired a safe and breathable atmosphere. It also acquired potable water and comfortable temperatures. Imagine that! Things were really developing nicely on this self-made world.

The eons rolled by, one after the other. Billions of years later, rainwater was flowing over some rocks, and it leached some minerals from the rocks. Those minerals found their way into rivers, streams, and oceans. By the luckiest of coincidences, the right combination of minerals came together and produced a soup that was rich in amino acids. It was in this primordial soup that the first living cell came into existence all by itself. In some versions of this fairy tale, a bolt of lightning struck the water, and jolted the cell to life. We are told in our schools and universities that all life on earth evolved from that first living cell.

In time, beautiful flowers, trees, vegetables, fruits, and other plants suddenly sprang up all over the earth. The varieties and colors were spectacular. All of these magnificent plants came into existence all by themselves. Not only were these plants gorgeous, but they were also very highly functional. They cleaned the

The Holy Bible: Is it Fact or Fiction?

air, provided oxygen, held the soil in place, and would soon serve as a source of nutritious food.

It's starting to look like this is part of a grand design, as if preparations are being made for some creatures that have not yet evolved, but that doesn't make any sense. After all, this is an evolutionary version of things where no designer is involved in the process. How could the random forces of nature possibly know what the future had in store and make preparations for it? I guess it all just happened without any particular rhyme or reason. Curious, isn't it?

If God didn't create fruits and vegetables with a specific purpose in mind, then they must have created themselves. So, instead of thanking God for our food when we eat, we should thank these wonderful foods directly. We should thank them for somehow anticipating what our nutritional needs would be, and then for having the foresight to create themselves for our benefit. If it sounds like pantheism to revere and worship all things in nature, so be it. I believe in giving credit where it is due. If the plants did all of this on their own, then lets thank them instead of God.

All forms of life, including plants, have a way of reproducing. Does anyone really believe that plants could have anticipated the need for reproduction and then arranged to evolve in a manner that would produce seeds for that purpose? That kind of preparation for the future required planning by an outside designer who had the ability to see the future. Let's stop kidding ourselves about what can happen all by itself. It defies logic and insults one's intelligence to claim that this all happened by itself.

If I were God, I'd be a little bit offended that I was not getting proper recognition for what I had done. One day, he will get that recognition, and a lot of people are going to be very surprised! On that day, there will be no more doubters, no more atheists, and no more evolutionists. At that time, everyone will be a believer.

Now, I would like to get back to our story and talk about that first living cell that accidentally

The Holy Bible: Is it Fact or Fiction?

created itself millions of years ago. Any time a story begins with the words "Once upon a time, millions of years ago . . .", watch out because you know another fairy tale is about to be told. Indeed, that's exactly what scientists have fabricated, and they are teaching it as if it were true. According to them, the 1.8 million different species of insects, birds, and animals, plus the billions of different micro-organisms that exist today, all came from that first living cell.

As that cell swam around in warm primordial waters, it branched out into countless varieties of marine animals, including fish, sharks, sting rays, whales, dolphins, eels, and a whole host of other creatures. Millions of years later, one of those sea creatures grew appendages and it quite miraculously developed the ability to crawl around and breathe air.

This newly-evolved creature slithered out of the water one day, and then some truly amazing things began to happen. This creature, which had lived its entire life in water, instantly adapted to life on dry land. It immediately developed the

The Real Origin of the Species

ability to breathe air, and it acquired the ability to move around and go from place to place. It boggles the mind how this creature survived for even 24 hours in this strange new environment, much less how it lived long enough to evolve into a higher form of life. Any other legless marine animal that happened to beach itself would have been dead within days, either from starvation or from being baked by the sun; but not this rugged little fellow. He was extremely lucky and very resourceful. The creature had previously eaten only foods that were available in the ocean, but it immediately found the perfect food supply on land that would nourish it for millions of years until it could evolve into another form of life. We know it lived for millions of years, because scientists tell us that evolution doesn't occur rapidly or perceptibly. It's a very slow and long, drawn-out process.

Just how far can a creature with no hands and no feet slither around in one day, and how much can it do for itself? Not very much, I assure you. Yet this amazing little guy somehow managed to slither its way around in that unfamiliar environment and find everything it needed. It found ample amounts of food every day. It found shelter from the sun. It also endured long periods of solitude (unless it was hermaphroditic and could produce more of its own kind without a mate). Are you buying any of this, so far?

At any rate, different branches of that animal developed into numerous kinds of insects, birds, and four-footed animals. At the risk of repeating myself, let me just say that it is truly amazing what can happen all by itself.

If you think that's amazing, it gets even better. After millions more years of evolution, different branches of those animals developed into bipedal beings, and began walking upright. Then, after branching off from different species of monkeys and apes, modern man finally evolved; and that's how the fairy tale ends. I hope you weren't shocked to hear that you came from apes instead of coming from the hand of God, but that's the way fairy tales are. Sometimes they have very scary endings.

The Real Origin of the Species

All life on earth supposedly came from that first living cell. That raises the question of where plant life came from. Plant and animal proteins are completely different, so they could not have come from the same source. This is one of those questions that no one is supposed to ask, and scientists just kind of gloss over it, hoping no one will notice that they don't have an answer. They haven't yet concocted a suitable fairy tale to explain where plants came from.

Isn't it amazing that all the events in this story happened all by themselves, with no designer and no planning? What's even more amazing is the fact that there are so many truly intelligent people who actually believe this story.

Did humans really evolve from monkeys?

The Holy Bible: Is it Fact or Fiction?

In the beginning, God created the Heavens and the Earth. . . . and God created man in his own image.

So there you have it! Those are the two main versions of our origins: creationism and evolution. One of these versions is an orderly, controlled, and purposeful process, and the other one is random and chaotic, and it relies very much on chance. One version is the truth, and the other one is an elaborate fabrication. Now that you have looked at both versions, it might be well to consider which one seems to be more plausible. Did God create all things, or did all things create themselves? Don't be afraid to challenge the conventional wisdom on this matter. Trust your instincts, and let common sense prevail.

Section II

Follow the Logic

If you were walking in a barren wasteland that had nothing but sand dunes for thousands of miles in every direction, you would logically and rightly conclude that this scene had been produced by the random and uncontrolled forces of nature. If you suddenly saw a house in the midst of that vast wasteland, you would not conclude that it, too, was produced by those same random forces of nature. Logic would tell you that a sentient being had designed and built it with a specific purpose in mind, and you would be right. It would be perfectly logical to come to the same conclusion when you look at the world and everything that's in it. Observe the order, beauty, harmony, and functionality of the world, plus the precise laws of physics by which it operates, and you can come to only one conclusion. Just follow the logic. Where there is order, harmony, and functionality on such a grand scale, there has to be a creator. If we follow this logic, it always leads us to God.

The Holy Bible: Is it Fact or Fiction?

I am sure that scientists would agree that there are no man-made objects in this world that did not first exist in someone's mind—take for instance such things as cars, televisions, appliances, computers, tall buildings, clothing, and so on. All of these things existed first in someone's mind before they came into existence in the material realm. If scientists can agree that this is true, it should be equally logical to conclude that something as magnificent as this world also first existed in the mind of God. Everything in this world was obviously created by someone for a specific purpose, much like the house in the midst of a barren wasteland. Objects come into existence only after they have been conceived in the mind. This is true on the human level, and I submit that it is also true in the grand scheme of things.

When humans conceive an idea for a new product, they bring it into existence with their hands, and it often takes years to accomplish. When God conceives an idea, he speaks and brings it into existence in mere moments. God is not bound by human limitations; if he were, he would not be God. We are created in the image of God, but that's where the similarity ends. There are vast differences between our powers and God's powers, and there are also vast differences between the way we operate and the way he operates. In view of those differences, it should not be difficult to grasp the idea that God could create the world in six days. The word "omnipotent" doesn't even come close to describing who God is. If you can't get your mind around the awesome and unlimited power of God, then your concept of God is much too small.

This means that you and I were not merely the random result of conception. We existed in the mind of God before we were even born. The Bible says that God knew Jeremiah even before he formed Jeremiah in his mother's womb. Is it any wonder then that God knows us so well, even down to the number of hairs on our heads? There are literally millions of possible outcomes from every act of conception, and every one of them is known to God in advance.

The fact that we were conceived in the mind of God before we were born has broad implications. It also means that when a baby is aborted, not only is a life aborted, but God's plans for that individual are also aborted. It's never a good idea to go against God's plans in order to pursue our own selfish ends. If you always choose to do the right thing, God will bless you. The joy and blessings that come to you will far outweigh any temporary benefit that you may get from going against God's wishes.

If we follow logic when trying to decide whether life evolved by accident or was created on purpose, I think the choice is clear. I urge you to follow the logic. Nothing exists that didn't first exist in someone's mind.

Section III

Is Evolution a Form of Idolatry?

Anything or anybody that claims to be the source of life is attempting to displace God, and that makes it a form of idolatry. Millions of people worship at the altar of evolution. Scientists are its high priests, and university professors are its chief proselytizers as they brainwash each generation of young impressionable minds. This form of idolatry is sanctioned and supported by the government. Many state and local governments require that evolution be taught in their schools, and they absolutely forbid the presentation of any alternative viewpoints. Constitutional experts should look into whether this constitutes state sponsorship of religion (a kind of "godless religion"). I question whether public schools and universities operated by the states should be using taxpayer dollars to teach a subject that is more like a religion than true science. There is nothing scientific about the Theory of Evolution, and that is why it is called a theory. Evolution does not even meet the minimum criteria to be called a true science. The theory is so illogical that it actually requires more faith to believe in evolution than it does to believe in God. I consider evolution to be a religion because it has to be accepted completely on faith. There is not one single shred

The Holy Bible: Is it Fact or Fiction?

of credible evidence to suggest that evolution is occurring or ever has occurred.

Scientists claim that they cannot accept the notion of creationism because God cannot be observed, verified, or duplicated in a laboratory, and true science is based on the ability to do these things. There is an inconsistency in how they apply that rule. If they are going to reject creationism on that basis, then they should reject evolution as well. There is no aspect of evolution that has ever been observed, verified, or duplicated in a laboratory. No missing links or transitional life forms have ever been found or observed because there are none. Evolution is accepted strictly on the basis of faith, and it should not be called science. It is no different than any other faith-based religion. It is, in fact, a kind of godless religion, one that any atheist could readily embrace.

It's very clever how Satan has duped the whole world into a subtle form of idolatry by dressing it in the mantle of science. I don't think it's an overstatement to say that evolution is idolatry. God has said, "Thou shalt have no other gods before me," and yet many people have done that very thing without even realizing it. When evolution is presented in the guise of a pseudo-science, it doesn't seem so terrible, and it isn't seen for what it really is. Terms like "evolution" and "natural selection" sound sophisticated and seem to be quite harmless. However, these words are nothing but cunningly disguised forms of idolatry because they put other forces in the place of God.

The so-called laws of physics and laws of nature are really the laws of God, and we do him a disservice when we ascribe these laws to some vague or nebulous entity like Mother Nature. It may sound cute, but it's a disservice nonetheless. God deserves credit and praise for what he has done, but the world seems determined to exclude him.

Most people who believe in evolution would recoil at the suggestion that they are really bowing down before an idol; but in a certain way, that's precisely what they are doing.

Section IV

The DNA Evidence is in, and a Positive Identification Has Been Made: God is the Real Father of Us All

Darwin's Theory of Evolution is full of flaws, and yet it is the foundation of modern education regarding the origin of life. Scientists continue to build on that shaky foundation, even in the face of new evidence that points toward the idea of a creator. Two such pieces of evidence are the discovery of DNA and the mapping of the human genome.

Genomes and DNA sequencing codes are the blueprint for all living things. They are the means by which cells are instructed to turn into a particular animal or plant. These codes direct the sequential formation of all the various body parts (brain, heart, lungs, liver, stomach, skin, bones, teeth, nerves, hormones, and immune system, just to name a few of the many thousands of other body parts). The human genome contains more than three million bits of information, all crammed into a microscopic package so small that it can't even be seen with the naked eye. It is truly a marvel of miniaturization, and yet the programming instructions that it contains are so numerous and so complex that they could not possibly have written themselves.

Each and every living organism has its own unique genome with similar amounts of coding. Imagine the vast amounts of programming it took to accomplish that, and yet we're supposed to believe it happened all by itself.

No one thinks that the programming that runs our computers could create itself. We instinctively know that something this complex could never create itself, and yet millions of people are perfectly willing to accept the notion that DNA coding somehow managed to produce itself. The human brain alone is more complex than our most advanced computers, and the DNA coding that produced the human brain is even more complex than the brain itself. Just as it would be illogical to assume that Microsoft Word wrote itself, it would be equally illogical to assume that DNA coding wrote itself.

We admire Bill Gates and give him a great amount of praise, honor, and money for his accomplishments in creating his amazing computer programs, and yet we give God nothing for his accomplishments, not even an acknowledgement of what he has done.

Several years ago, the scientific community decided to create a map of the human genome in order to better understand how it operates. The human genome is, after all, the operating system for humans, much like the operating system in a computer. One researcher who was involved in this project was so impressed with the beauty and complexity of the human genome that he referred to it as "the language of God".

The Real Origin of the Species

A very tiny portion of the human genome:
Is it the language of God?

It may not exactly be the language of God, but it certainly is his handiwork. It's hard to believe that any scientist could look at the sophisticated structure of the human genome and conclude that it all happened by some evolutionary accident. The design is purpose-driven and is clearly the work of a superior intelligence.

According to scientists, evolution is an uncontrolled and persistent process. It never stops. If the human genome had evolved randomly and by accident, it would have continued to evolve, producing one variation after another, probably in some sort of mutated form. Instead, the human genome has remained fixed, unchanged, and totally reliable. All other plant and animal genomes have also remained unchanged, just as the designer intended. A number of animals have become extinct, but none have changed into something different. Today we still see the very same animals that have existed since the beginning of time.

There has not been even the slightest deviation in any of the genomes, and there is no evidence that any kind of evolutionary process is occurring. When God was creating the world, he announced at the end of each day's work that what he had made was "very good". It was perfect just the way it was, with no need for any kind of evolutionary tweaking to get it just right. The Bible says that God does not change. Scientists can look for changes all they want, but they will never find any.

Darwin did not know about genomes and DNA when he wrote his famous theory. If he had, I think even he would have altered his conclusions about the origins of life; but scientists steadfastly refuse to abandon the Darwinian model.

Scientists are so desperate to find examples of evolution that they now claim to have observed evolution taking place among micro-organisms. It's a process they call micro-evolution. I am convinced that what they are seeing under the microscope is not really evolution. Those micro-organisms are not changing into new and unpredictable life forms.

The processes that scientists refer to as micro-evolution are probably nothing more than hybridization, metamorphosis, or mutation. Under hybridization, one kind of organism that has limited compatibility with a different kind of or-

The Holy Bible: Is it Fact or Fiction?

ganism can mate and produce a totally different kind of organism. An example of hybridization is when a horse and a donkey are mated and produce a mule, which is a totally different animal than either of its parents. Another example of hybridization is when a plum is crossed with an apricot to produce a totally different kind of fruit called a pluot or plumcot. Hybridization occurs frequently among plants.

Metamorphosis is a multi-stage process that is pre-ordained by the Creator, and the outcome is totally predictable. An egg morphs into a caterpillar. After a period of time, the caterpillar is transformed into a moth or a butterfly.

Mutations occur when DNA is damaged by nutritional deficiencies, physical trauma, exposure to chemicals, or some other outside agent. Mutations usually produce grotesque variations of the original. They do not produce anything that is new or better than the original.

Whatever process it is that scientists are observing under the microscope, we can be certain that

The Real Origin of the Species

it is operating within the boundaries and laws established by the Creator. The outcomes are more or less predictable, and no dramatically new life forms are evolving in an uncontrolled manner.

Another specious claim made by scientists is that birds evolved from dinosaurs. They make that claim based on the fact that both classes of animals lay eggs, walk on two legs, and have skeletons that are similar. They also say that embryonic chickens have tails like dinosaurs, but those tails drop off before the chicks are hatched. The claim that birds evolved from dinosaurs is an absurd assumption. It is the same kind of assumption on which they base their claim that man evolved from monkeys and apes. The similarity between these life forms was simply a matter of divine expediency. God created a basic chassis on which he could build many different life forms. It was not necessary for him to go back to the drawing board and come up with a totally new chassis every time he made a new creature. It is much the same as when a car company designs one basic chassis, and then

The Holy Bible: Is it Fact or Fiction?

puts several different bodies on the same framework. There may be as many as three or four different models mounted on the same chassis or skeletal system.

The number of dissimilarities between dinosaurs and birds is far greater than the similarities. One animal is gigantic in size, and the other is comparatively small; one weighs many tons while the other weighs only a few ounces or pounds at most; one has very dense bones while the other has a partially hollow skeleton to keep its weight low enough for flight; one has wings and can fly, and the other one has no wings and is earthbound; one has teeth and no beak while the other one has a beak and no teeth; one has feathers while the other one is featherless—the list of dissimilarities goes on and on. Although some of the smaller dinosaurs were able to fly, there is no evolutionary link between them and the wide variety of birds we see today. Dinosaurs could never turn into birds, not even in the millions of years that scientists say it took for them to do so. It takes a lot of imagination and a great leap of faith to make a connection between dinosaurs and birds. For scientists to single out a few similarities while ignoring all of the dissimilarities is quite a stretch. It shows the extremes to which they will go in order to make the case for evolution and keep God out of the picture.

Scientists are continually building new and more powerful telescopes with which to probe the universe. They readily admit that they are hoping to find life on other planets. During a radio interview, one scientist recently stated that he believes there is another purpose behind the relentless effort to find life on other planets. That purpose is to show that life happens randomly and spontaneously throughout the universe and that there is nothing special or exceptional about life on earth. They will do anything to keep from acknowledging God's involvement in the creation of this world and its inhabitants. So far, they have found no evidence that life, as we know it, exists anywhere else in the universe.

I'm sorry to inform the scientific community that life on earth is indeed special and unique, the singular product of a master designer. It is the object of God's attention and supreme affection. God is very much involved with this world in spite of the best efforts of scientists to dismiss him.

Evolution is not only improbable—it's utterly impossible. It's impossible because it takes an enormous amount of energy and strength to evolve into something that is new and better than the original. From the moment of their birth, all living things begin the inevitable process of dying. So how can anything that is dying and steadily losing strength possibly evolve into something completely new? The fact is, it can't happen. It never has, and it never will.

Section V

Positive Proof of the Veracity of the Bible:
A Look at Prophecy and History

The story of Creation is just the lead story in the Bible. There are many other fantastic stories

The Real Origin of the Species

in its pages, as well. Many of these stories are so amazing that it doesn't seem that they could possibly be true. For example, who could believe that God once punished mankind for its sins by sending a flood that covered the entire world, and that only one family survived along with a boatload of animals? What rational person could possibly believe that the Red Sea once parted to let the nation of Israel cross over on dry land, and then the sea closed back up again to drown the Egyptian army that was pursuing them? Those kinds of things didn't really happen, or did they?

Then there are stories like the one about a teenage boy who defeats a giant by using nothing more than a slingshot; or the one about a man with superhuman strength who single-handedly defeated one thousand attackers by using nothing more than the jaw-bone of a donkey as a weapon. There are stories about a man surviving a whole night in a den full of hungry lions and about three young men being thrown into a fiery furnace and coming out completely unharmed. There is also the story of a man who survived after spending three days and nights in the belly of a whale.

If you think those stories are hard to believe, how about the one where a virgin gives birth to a boy who calls himself the Son of God, and after he reaches adulthood, walks on water, turns some water into wine, restores sight to the blind, makes the lame to walk again, heals incurable diseases, and even raises people from the dead? As if that weren't enough, he dies a horrible death in order to save mankind, is brought back to life after three days, and then ascends into heaven. Pretty fantastic stuff, isn't it?

Can these stories possibly be true? The Bible says of itself that "all scripture is given by inspiration from God" (2 Tim. 3:16), and that holy men spoke as they were moved by God. These statements and the stories described in this book are either true or they are false, and just think of the implications if they are true! I want to provide you with a solid basis for believing that they are indeed true.

The Real Origin of the Species

In order to accept the Theory of Evolution, it is necessary to shut down the logic centers of your brain, and then make a blind leap of faith. No such blind leap of faith is necessary in order for one to believe that the Bible is true. There is a prophetic and historical basis for believing in the Bible. One sure test of the veracity of a prophecy is whether the events that were predicted actually occurred, and can be confirmed by historical and geological records. The Bible is replete with prophecies that have been fulfilled in every detail. In one of them, God told Noah that there would be a worldwide flood 120 years before it actually happened. Evidence for the Flood is presented in chapters 2 and 3 of this book. Some other evidence of the Flood is found in the many fossils located on mountaintops in South America. Among those fossils are huge quantities of clams, oysters, and other shellfish, which makes it clear that those mountaintops were once under water, just as it says in the Bible.

In the Bible, the book of Daniel contains two separate prophecies that predicted the rise and fall of the Persian, Grecian, and Roman empires hundreds of years before those events took place. Specific characteristics of these empires were described in great detail by the prophet Daniel. One of those prophecies even foretold the division of Europe into ten separate countries after the fall of Rome. It's thrilling to read these prophecies and see how accurately they were fulfilled. Only God could describe the future so accurately and in such great detail.

One event that was predicted more than any other event in history was the coming of the Messiah. There are approximately 40 prophecies in the Bible pertaining to the Messiah, but Jewish scholars say that the ancient scrolls contain about 300 prophecies that foretold the circumstances of his birth, including the location, the approximate date, and the fact that he would be born of a virgin. Those prophecies also foretold his life, his physical appearance, his ministry, his sufferings, and his death.

The Real Origin of the Species

How can we be certain that Jesus Christ is the one referred to in those prophecies? We do that by matching the prophecies with events in history.

A prophecy in the book of Daniel says that a prince would arise who would destroy the city of Jerusalem and the temple <u>after the Messiah is cut off</u> (the underlined words are crucial in identifying Jesus Christ as the Messiah). History shows that it was Titus, a son of Augustus Caesar, who was the prince identified in the prophecy. He was in charge of the army that destroyed the city and the temple. We know that the destruction of Jerusalem occurred in 70 AD. According to the prophecy, that means the Messiah had already come and been "cut off" or crucified by 70 AD.

The only person who came along at the right time and met the requirements of all 300 prophecies was Jesus Christ. Any statistician will tell you that the odds of one person fulfilling all of those prophecies are in the trillions. The following example will give you some idea of the odds. Imagine that there is a huge field as big as the state of Texas, and that this field is covered with silver dollars stacked one foot high. If only one of those silver dollars was marked with an "X" and then hidden somewhere in that field, your odds of finding that one coin with the "X" on it would be as great as the odds of one person fulfilling all 300 prophecies. So it's clear that Jesus Christ is the one whose life was foretold by the prophets.

Section VI

Implications and Recommendations

The evidence that I have provided to support the veracity of the Bible is only a small portion of what is available. I hope I have presented enough to be convincing and to whet your appetite for more knowledge. Now that you have seen the evidence, you have to decide what to do about it, and there are only two choices: (1) you can ignore or reject it, go on with your life as usual, and miss out on one of the greatest experiences of your life; or (2) you can come to the conclusion that the Bible is, in fact, true,

The Real Origin of the Species

and then explore where that knowledge will lead you.

What are the implications if you do accept the veracity of the Bible and decide to go all the way and follow this newfound knowledge where it leads you? If you go so far as to acknowledge that Jesus is the Messiah and to accept him as your Savior, the implications for you will be numerous and life-changing. First of all, let me congratulate you for having the courage and wisdom to reject the erroneous teachings of evolution. To use the Christian vernacular, you have been born again, and you are a new creature in Christ Jesus. This is only the first of many wonderful adventures in faith that you will have, and you will be amazed and delighted at what God will do in your life when you partner with him and live in harmony with his will.

Among the many changes you can expect are the following:

1. You will attain a new and higher level of consciousness, one that will give you a

The Holy Bible: Is it Fact or Fiction?

keener awareness of God's presence and involvement in your life.

2. You can now enter into a very real and personal relationship with Jesus Christ, a relationship that he wants to have with you. He says, "Behold, I stand at the door [of your heart], and knock: if any man hear my voice, and open the door, I will come in to him, and will sup [or dine] with him, and he with me" (Rev. 3:20). Those are not just idle words spoken by someone who has no intention of keeping his promise. If you open the door to him, he will come in.

You will notice that he knocks at the door, and waits for the door to be opened. He will not force his way into your life. He enters only when he is invited to do so. I urge you to invite him into your life, because a one-to-one interaction with the Messiah (the one sent from God) is an experience you don't want to forfeit by refusing to open the door. Just as Jesus Christ calmed the angry seas and quieted the raging storms when he was here on earth, he will bring calm and peace to your life as well.

The insights I have shared with you in this book are the result of Jesus having supped with me over the course of my life. Let him sup with you, and see what wonderful things he reveals to you.

3. You will come to know that sin separates you from God, and the end result of sin is misery and death. The law of God says that sin is to be punished by death ["the soul that sinneth, it shall die" (Eze. 18:4)]. Jesus realized that since we all have sinned, we would all die and be eternally separated from God. He loves us so much that he volunteered to die in our place. By his sacrifice, the requirements of the law could be satisfied, and God could avoid losing us, his precious and beloved creation. It does wonders for a person's sense of security and self-esteem to know that we are loved so much by one so powerful. We can never be truly happy and fulfilled until our relationship with God

is restored; and if you have accepted Jesus Christ as your Savior, you've taken the first step in restoring that relationship.

4. You will acquire an intense desire to discover and do God's will. The best way to satisfy that desire is to read a portion of the Bible every day. If you can find the time to watch television for three or four hours each day, you can surely devote 30 minutes a day to the study of God's Word.

I recommend that you begin by reading the four gospels (Matthew, Mark, Luke, and John) and the book of Acts. After that, I recommend a quick reading of the entire Bible just to give you an overview of what it contains. Then you may wish to go at a slower pace and study specific topics or your favorite portions of the Bible.

The use of Bible commentaries written by biblical scholars may facilitate your understanding of the more difficult portions of the Bible. Some of the best commentaries that

I know of are in a series of books written by Ellen G. White, and they include the following titles: *Patriarchs and Prophets, Prophets and Kings, Steps to Christ, Desire of Ages, Acts of the Apostles, Daniel and Revelation,* and *The Great Controversy.* These books may be obtained by calling 1-800-325-8492. Other Bible commentaries may be found at your local Christian bookstore.

The study of God's Word should be pursued throughout the remainder of your life. You will sharpen and expand your mind by this exercise, and you will learn something new every time you read it.

5. In addition to reading your Bible regularly, it is equally important to talk to God every day in prayer. As you read the Bible, you will discover that those people who had the closest relationship with God also had the most active prayer life. Some of them prayed as many as three times a day. If you want to have a dynamic relationship with God, follow their example.

When you pray, talk to God just as you would when speaking to a friend. Not only does God hear your prayers, but he delights to answer them.

Jesus taught us how to get answers to our prayers. He said that when you pray, believe that you already have the things for which you are praying, and you will have them. You will notice that he doesn't say to believe that you will have them at some vague, indefinite time in the future. He says to believe at the very time you are praying that you already have what you are asking for.

I read that text many times over the years before I realized what Jesus was trying to tell us. Do you recall when I said in a previous chapter that nothing exists in this world that didn't first exist in someone's mind? That same principle applies when you pray. If you can get the things you are praying for to first exist in your mind by faith, then they will soon come into existence in the material world. You have to own it mentally before you can receive it physically. The Bible says in Hebrews 11:1 that "faith is the substance of things hoped for, the evidence of things not seen". Faith makes things real in the spiritual dimension before they become real in the material dimension.

Jesus said to many of the people he healed, "Thy faith hath made thee whole". Faith sets in motion the forces that will bring your desires into existence. Christ also said that it doesn't even require much faith to get your prayers answered. He said that if you only have faith the size of a mustard seed, you will have what you ask for.

Your faith will grow stronger day-by-day as you see for yourself that God does indeed answer the prayer of faith.

6. You may desire to associate with others who share your newfound faith. You would do well to search for a Christian church that appeals to you. Jewish converts may join a Christian church, or they may prefer to lo-

cate the nearest congregation of messianic Jews. If you need assistance in locating a church near you, call 301-680-6000.

I don't recommend that you make your journey alone. Some sort of fellowship with like-minded individuals will strengthen your faith and speed your spiritual growth.

7. With your enhanced vision and heightened state of consciousness, you will see things in a totally different way. For example, the next time you eat an apple or any other kind of food, you will not take it for granted. Instead, you will realize that it was designed by someone who had your nutritional needs and your personal tastes in mind—someone who then put it in the heart of farmers to grow it, truckers to deliver it, and merchants to sell it, all so you could enjoy it. It will infuse you with a profound sense of gratitude.

God knew that you could never supply all of your needs and desires by yourself, so he put into place a network of fellow human beings who all contribute in some way to meeting your needs. A few of the many examples are the people who cut or coif your hair; deliver your mail and packages; serve the food at your favorite restaurant; remove your trash; clean your streets and buildings; defend or govern your country; design and manufacture the clothing, cars, computers, televisions, and other appliances you use; care for you when you are sick; build your house, etc.

Before your transformation, you might have viewed all of this commercial activity as nothing more than normal market forces at work. Now you know that all of these people are intermediaries through whom God works to meet your particular needs. That thought should give you a new appreciation for your fellow man.

8. Until now, you may have assumed that all of your possessions and all of your success in life were due solely to your intelligence, ingenuity, and hard work. Those things are

only a small part of it. Now you know the truth. The Bible says that it is God who gives you the power to obtain wealth. God knows you so intimately that he can determine ahead of time just what opportunities, experiences, and people to bring your way in order to advance your well-being and happiness. God works unceasingly on our behalf, and he deserves credit and gratitude for all that he does.

9. God is so eager to be involved in our lives, that you will be amazed at how quickly the answers to questions and solutions to problems will come to you, often before you even have time to ask for his help. Now that you have been transformed, you will know where your help comes from.

10. Now when you look at the wonders of nature; hear a bird sing or see one soar in the sky; gaze upon the infinite varieties and colors of flowers; watch the ocean waves roll in on the beach; or behold the beauty of a sunrise or sunset, your amazement and

gratitude will overflow. You'll be moved to exclaim, "Oh Lord, my God, how great and wonderful you are."

11. There are many wonderful promises in the Bible. Now that you know the Bible is true, you can immerse yourself in those promises and embrace them fully, knowing that they were made by one who can and will deliver on each one of them. These promises give hope and encouragement, especially in difficult times.

    God even challenges us to test his promises to see if they are true. In Malachi 3:10, he says to obey him by giving him one-tenth of our income, and then He says, "Prove me now herewith, saith the Lord of hosts, if I will not open you the windows of heaven, and pour you out a blessing, that there shall not be room enough to receive it." That's a very explicit promise, and one that's easy to prove or disprove. I have heard countless testimonies of people who have tested this promise and have been wonderfully blessed in many different ways. God is faithful, and keeps every one of his promises.

I began proving the Lord's promises when I was 16 years old; and now, 56 years later, I am able to confirm unequivocally that God's word is true and he keeps his promises. I have been blessed beyond measure. All of my needs have been abundantly supplied, and most of my desires have been realized. I have been amazed to watch God open doors of opportunity for me at just the right time in my life. I encourage you to accept the Lord's challenge, and to begin testing his promises for yourself.

One of the most glorious of all promises in the Bible is the promise that Jesus is coming again. When he returns, he will bring an end to sin, suffering, disease, and death. Those who have been faithful and obedient will spend all eternity in his presence and in the presence of our heavenly Father.

If the prospect of spending eternity with

your creator doesn't send a thrill through your soul and set your heart to racing, you'd better have someone check your vital signs. You might be dead already and just not know it. This is not "pie in the sky", folks. This is a real event, and you do want to be ready for it.

Christ's return may be sooner than most people realize. When Jesus was on earth 2,000 years ago, he said that there would be numerous earthquakes around the world just before he comes back. While this book was being written, there was a dramatic increase in seismic activity all over the globe. Within less than two months, powerful earthquakes struck such places as Haiti, Chile, Taiwan, Japan, and Turkey. Soon after that series of earthquakes, additional quakes struck Chicago, Illinois and southern California. Then Chile was hit once again. We were all so focused on those earthquakes that most of us didn't realize that there were 20 other minor earthquakes in different places during that same time period.

A short time after those earthquakes ended, a new series began. California was jolted once again, and that was followed by a quake in Mexico. Much to everyone's surprise, earthquakes were felt in places that don't normally have them. Tremors were felt across the entire northeast from Vermont to Michigan and from Maryland to Canada.

We've had earthquakes before, but I think there is something different this time. It could be that Christ's prophecy is being fulfilled right before our very eyes. This could be a sign that he is soon to return. We can dismiss it as just another natural phenomenon, and go on with life as usual; or we can take it seriously and get our hearts ready for his return.

Christ will return one day surrounded by millions of angels. His return is going to be the most glorious and spectacular event in earth's history, and I hope this book has prepared you for it.

The Real Origin of the Species

Artist: Robert Ayers. Used by permission of Review and Herald Publishing.

Good News! Jesus Christ is Coming Again

My goal in writing this book has been to turn the hearts and minds of people back to God, and to provide them with a solid basis for their faith. I hope I have adequately made the case that the Bible is true when it says that God in heaven is our Father, and that he alone is the *real* origin of the species.

Reader Survey

Please complete this survey form to help us determine the effectiveness of this book. After you have filled it out, please remove this page, put it in an envelope, and send it to:

> Truth Above All Ministries
> Attn: Oscar J. Daniels, Sr., PMB 201
> 14625 Baltimore Avenue
> Laurel, MD 20707-4902

1. Gender:
    Male ☐   Female ☐

2. Age Group:
    Under 20 ☐   20-40 ☐   Over 40 ☐

3. Education:
    High school or less ☐
    One or more years of college ☐

4. Before reading this book, what did you believe:
    Creationism ☐   Evolution ☐

Next, complete either question 5 or 6, as applicable.

5. If you were previously a creationist, has your faith been strengthened?
    Yes ☐   No ☐

6. If you were previously an evolutionist, have your views changed?
    Yes ☐   No ☐

If this book has led you to Christ, or if it has strengthened your faith, please complete the Reader Survey form on the previous page of this book. Also, please share this book with someone else.

We invite you to visit the following blog site for additional information that may be of interest to you:
http://truthaboveallministries.blogspot.com

To obtain additional copies of this book, visit http://www.LNFBooks.com.

To receive a response to your questions and comments, send an e-mail to taaministries@gmail.com, or write to Truth Above All Ministries, c/o Oscar Daniels, 14625 Baltimore Avenue, PMB 201, Laurel, MD 20707-4902.

We invite you to view the complete
selection of titles we publish at:

www.LNFBooks.com

or write or email us your praises,
reactions, or thoughts about this
or any other book we publish at:

TEACH Services, Inc.
P.O. Box 954
Ringgold, GA 30736

info@TEACHServices.com